This book Belongs to

_____

_____

_____

Copyright©2023: Published in the United States by Krystle Duncombe.

No part of this book may be reproduced or used in any manner without the prior written permission of the copyright owner.

Tracing numbers

Trace the numbers and words below.

2

two

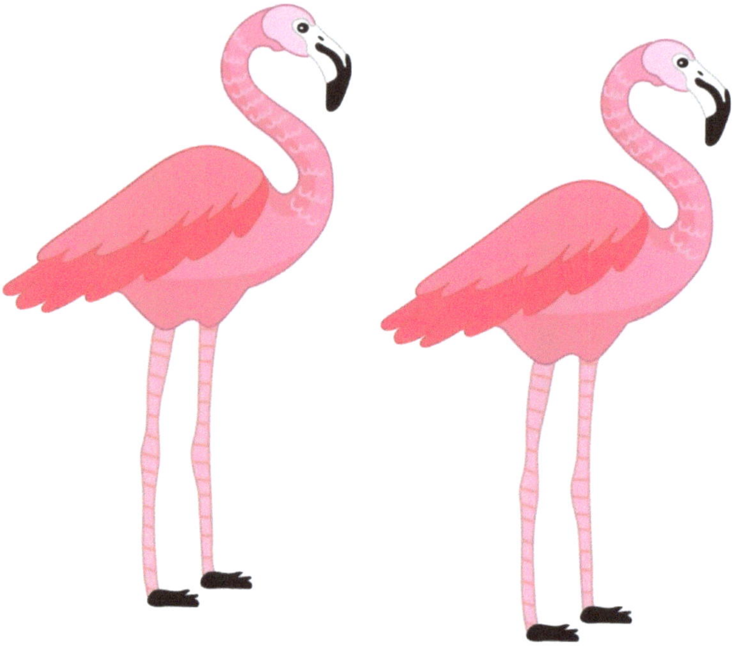

2 2 2 2 2 2 2 2

2 2 2 2 2 2 2 2

two two two

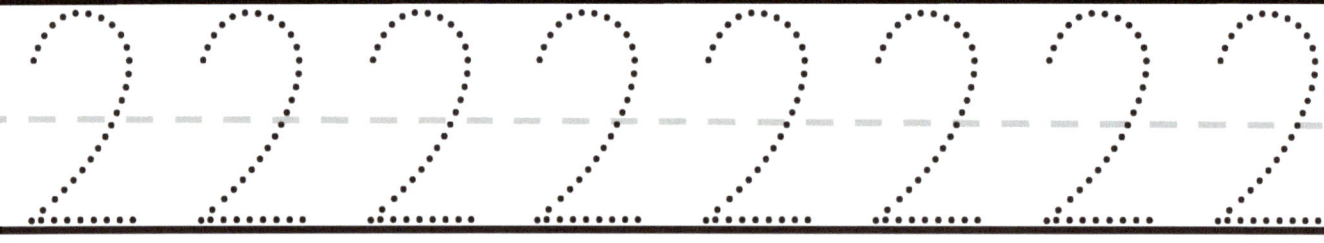

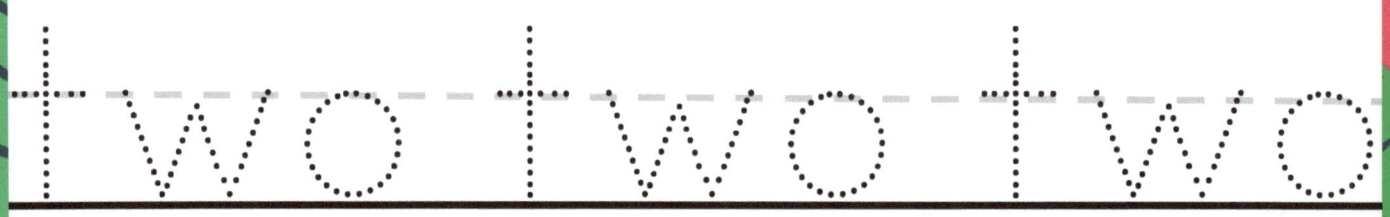

Tracing numbers

Trace the numbers and words below.

Tracing numbers

Trace the numbers and words below:

four

Tracing numbers

Trace the numbers and words below:

Tracing numbers

Trace the numbers and words below.

6
Six

6 6 6 6 6 6 6 6

6 6 6 6 6 6 6 6

six six six

Tracing numbers

Trace the numbers and words below.

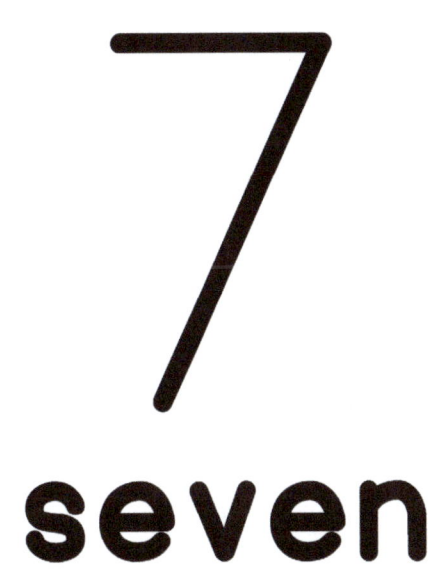

Tracing numbers

Trace the numbers and words below.

Tracing numbers

Trace the numbers and words below.

11
eleven

Tracing numbers
Trace the numbers and words below:

↓1 ↗2

twelve

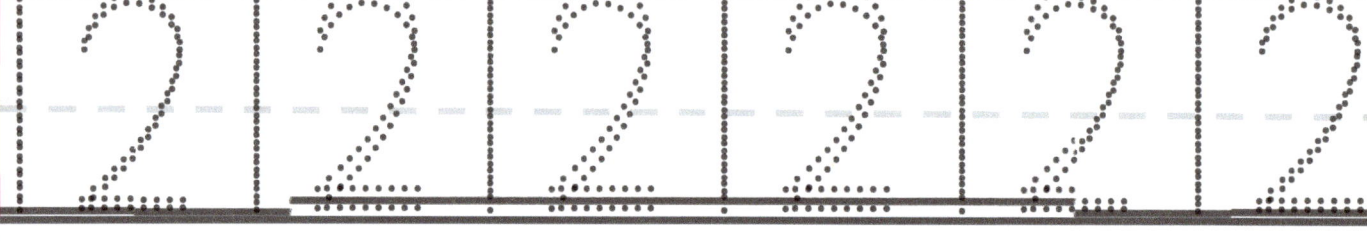

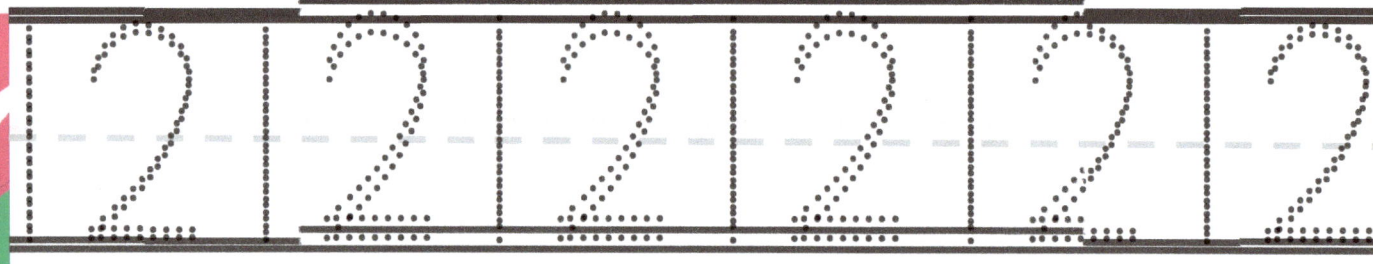

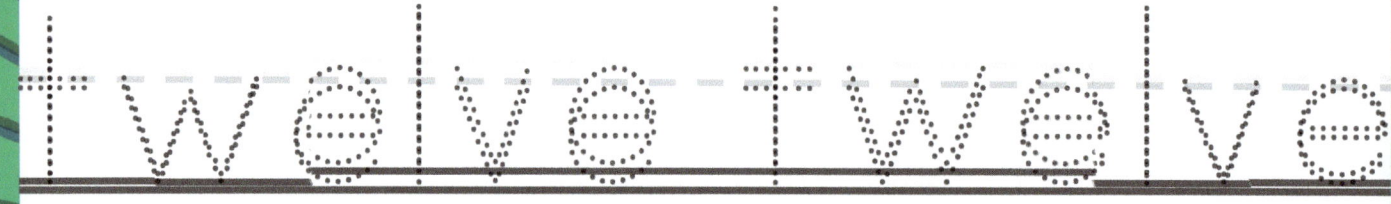

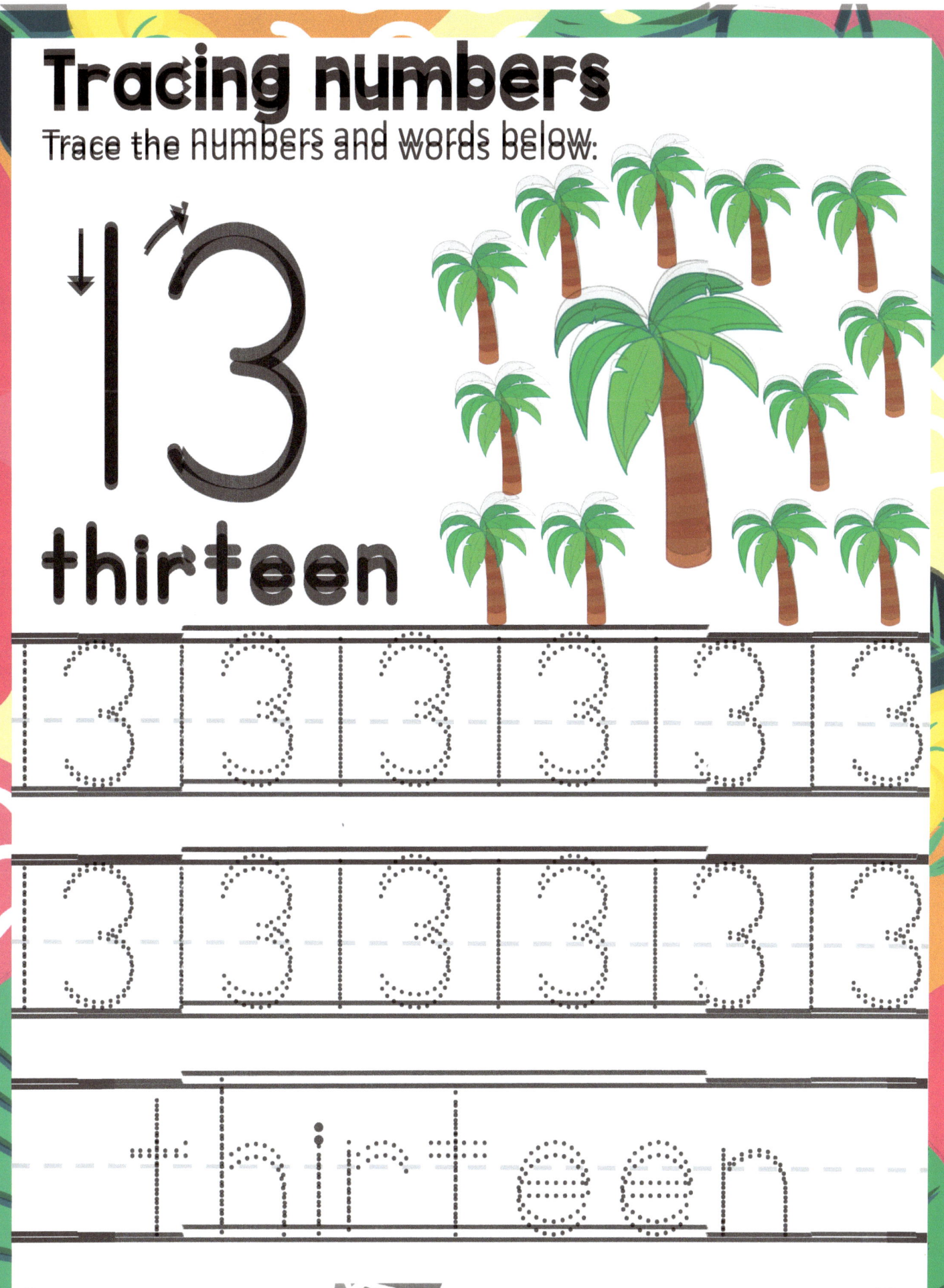

Tracing numbers

Trace the numbers and words below.

Tracing numbers

Trace the numbers and words below.

Coloring time!

Coloring time!

Coloring time!

Coloring time!

Coloring time!

Coloring time!

Coloring time!

Coloring time!

Coloring time!

Coloring time!

www.ingramcontent.com/pod-product-compliance
Lightning Source LLC
Chambersburg PA
CBHW051952210526
45473CB00023B/1870